The Mechanics of Clockwork

Lever Escapements,
Cylinder Escapements,
Verge Escapements,
Shockproof Escapements
and Their Maintenance
and Repair

British Library Cataloguing-in-Publication Data
A catalogue record for this book is available from
the British Library

Contents

THE ESCAPEMENT

ALL of the movement beyond the escape wheel pinion is
called the escapement, and here the heart of the subject is
reached. It is obvious that a watch with a poor power supply,
a roughly finished train, and motion work thrown together,
will not give good results, however fine the escapement. It
is also obvious that unless the escapement is properly made
and adjusted, all the other money and effort spent on the
watch are thrown away.

Only five main types of watch escapement have occupied
the attention of horologists since watches were first made :
many others may have been invented and re-invented, but

only the verge, cylinder, lever, duplex, and chronometer have been used in any numbers. Of these the lever is the most overwhelmingly popular because of the excellent results it gives on the small size movements of to-day. The lever has settled down to one theoretical shape although a technical specialist has published notes on fifty variations.

The verge is now entirely discarded, although verge watches still exist and even continue to go. It was an adaptation of the foliot balance used for the clocks of the sixteenth century. Wear on the train was excessive and good timekeeping was impossible ; even the best verge watch could not do better than keep time within ten minutes a day. More often than not the error was thirty minutes.

The duplex escapement was at one time used on many expensive watches and on many thousands of a cheap type of watch not now in production. Briefly explained, the escape wheel had two sets of teeth, one set horizontal and the other vertical on the rim of the wheel. A small roller with a notch to allow an escape tooth to pass and finger to take the impluse were fitted to the balance staff. It was a " frictional rest " escapement, the balance was never free, and as such unreliable. It had the knack of missing a beat when jolted. The duplex watch is probably more rare to-day than the old verge in the antique shops.

The escapement to become very popular on its introduction (anything to replace the verge was sure of a ready welcome) was the cylinder. This is still made in considerable numbers, in millions a year, in Switzerland, and has its followers as an escapement suitable for a cheap watch. Its use is declining, the lever can be made so much more reliable.

The disadvantages of the cylinder escapement are the " frictional rest " (one part rubbing another under pressure), when the escape tooth is held against the cylinder wall, and the lack of free action of the balance. The balance must be light, and is seldom compensated. Adjustment to position errors cannot be made. Very early attempts were made to reduce friction losses by making the cylinder of ruby or sapphire ; many watches so fitted exist and work to-day.

For very cheap movements the pin-pallet, a simplified form of lever escapement, is regarded as satisfactory. It can be made with stout parts and plenty of clearance, so that with a strong mainspring it can go and keep going. It is the alarm clock escapement adapted to a watch. In America watches of this type are known as pocket clocks, an apt and not derogatory description.

The chronometer escapement is the most precise in performance, but as it is very expensive to make, and delicate to use, is not in favour for watches. Nevertheless many fine chronometer escapement watches continue to be made ; when carried and used with the greatest care nothing is better. The most use made of the escapement is for ships chronometers, these clocks being fitted in a swinging mount so that the movement of the ship will not disturb them. Their purpose is to enable the seamen to discover his longtitude by comparing sun time at his position with Greenwich mean time as given by his chronometer.

To say that eighty-five per cent. of the watches by number and ninety-five per cent. by value embody the lever escapement would be to make a knowledgable guess. Exact figures are not available. There are many retail horologists who do not stock cylinder watches and many who beg to be excused repairing them so strongly do they advocate the dominating superiority of the lever.

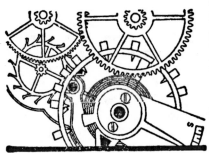

The Lever Escapement.
(See also page 32.)

It has been explained previously that the heavy pressure of the mainspring has been somewhat controlled and translated into speed by the gearing up of the train. From the fourth wheel this quick circular motion is received by the escape wheel pinion and still further speeded up by the larger revolving circumference of the escape wheel teeth. Here a change takes place. From being a rotating action it is translated into reciprocation and transferred to the balance in alternating directions. The power of the mainspring is thus allowed to drip away—to escape, hence the term escapement —in a steady stream which is more or less under precise

3

control. The power expends itself in swinging the balance by flexing and re-flexing the balance spring.

Important points in the escapement are the tooth shapes of the escape wheel and the exact shapes and angles of setting of the pallet stones. In detail, the lever, something like an inverted anchor (some old models were known as anchor escapements) is pivoted centrally and as nearly balanced as possible. The inward end is pincer shaped, the pincer ends being the pallet jewels ; the outer end carries a small fork and a projecting guard pin. On the balance staff are two rollers, one fitted with a jewel stone (known as the ruby-pin without regard to the material from which it is made), the other has a notch which allows the guard pin under the fork to pass.

The lever escapement, its principles of design, the exact angles and dimensions of its various components, its precise adjustment and functioning, are fruitful subjects for discussion in the industry. Essays with elaborate and intricate drawings and learned lectures followed by heated discussion have been made public, and there will be many more. The very excellence of the escapement attracts close study and long argument. Some think it is now near perfection, others say it is not and never will be. In this strife, whatever happens to the strivers, the subject of the argument gains.

The circular form of the escape wheel is of the first importance ; the club-shaped teeth must be all exactly formed. This might be easy if each wheel could be made with care and every operation closely watched. Made in that manner there would not be enough watches to go round and the price of the few would be ridiculously high. The wheels must be made by automatic machines so adjusted that each is turned out ready for inspection and use without tinkering.

Escape Wheel.

The lever pallets are known as the entry pallet and the exit pallet. The entering pallet is naturally the first to function. The heel of an escape wheel tooth will hit the entry pallet a mere fraction above the corner, enough to lock the wheel while the balance completes its swing and returns. On the return swing unlocking takes place and at that moment, whilst the ruby jewel is in the lever fork, the face of the escape wheel tooth passes over the impulse face of the pallet, giving it a push outwards, which the shape and angles of the lever translate into a sideways push on the ruby pin. Pushing out the entry pallet brings in the exit pallet and the tooth next but one ahead locks on it when the cycle is repeated. The complete cycle of operations of lock draw and impulse takes place 300 times a minute.

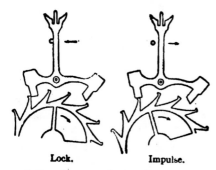

Lock. Impulse.

Action of escape wheel tooth on pallet in lever escapement.

The many millions of times this happens in a year can be calculated by a simple multiplication table. When the total is arrived at—it is many millions—it becomes incomprehensible how such tiny pieces of metal and jewel can withstand continuous and heavy wear. The secret is perfection of manufacture and a minute smear of oil. Because of the many moving parts about the pallet stones two small holes are usually drilled in the bottom plate through which the action of the pallets can be observed and a little oil applied.

The pivots of the lever are the first in the watch to turn from side to side, instead of rotating ; the next, and last, are the balance staff pivots.

The power from the mainspring of the watch having been transferred by the revolving escape wheel to the reciprocating lever at the pallets, is again transferred by the fork at the opposite end of the lever to the balance spring. Here it is stored momentarily in the effort of winding up the balance spring and is finally dissipated in the movement of the balance wheel, which is the flywheel of the watch.

Lever Escapement.—This escapement, the most important of all, is now found in nine out of every ten watches made. It may be in various forms, but all are made on one principle. Fig. 120 shows the form used in English watches; it is of the

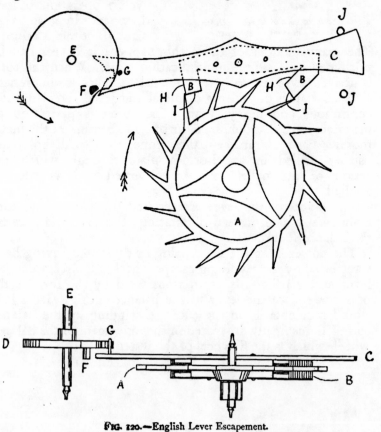

FIG. 120.—English Lever Escapement.

6

right-angled variety, and has pointed or "ratchet" teeth. Other forms are shown in less detail; the " straight line," in Fig. 121. In this the scape wheel, pallets, and balance are all planted in a straight line. Most Swiss and American watches are made thus. This figure also shows a "club-tooth" scape wheel, used in most foreign watches. The " pin-pallet " escapement (Fig. 122) has round pins for pallets, and the inclines are on

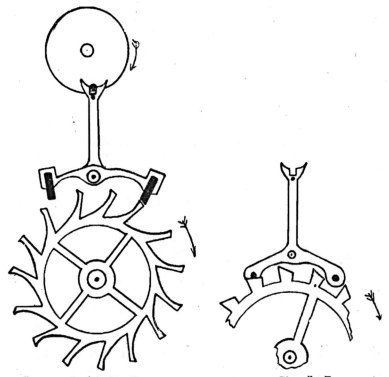

Fig. 121.—Straight Line Escapement.   Fig. 122.—Pin-pallet Escapement.

the scape teeth. The " club roller " (Fig. 123) and the "rack lever" (Fig. 124) are both old forms, now discarded, but still occasionally found in old watches.

In the rack lever the balance was never free. It was always moving the lever and pallets, and upon them rested the scape-wheel teeth. The modern forms leave the balance free, except while unlocking and receiving impulse. This was the origin of the words "detached lever" engraved upon old watches. It was to distinguish them from the rack lever, which was not "detached."

Referring to Fig. 120, the lever escapement consists of a fifteen-toothed scape wheel, A; a pair of anchor-shaped

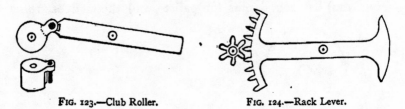

FIG. 123.—Club Roller.   FIG. 124.—Rack Lever.

pallets, B, fixed rigidly to a lever, C; and a roller, D, fixed to the axis of the balance E. The power of the mainspring is transmitted through the train wheels to the scape wheel. The teeth of this wheel are made to deliver impulses to the lever by passing in turn across the inclined faces I of the pallets. The lever passes these impulses on to the balance by means of a ruby pin, F, set upright in the roller. Between the impulses the balance spins freely, and the scape wheel is "locked" by the points of the teeth resting upon the "locking faces," H, of the pallets.

The action is as follows: In the figure one tooth of the scape wheel is resting on the locking face H of one pallet. The balance and roller D are returning in the direction shown by the arrow. When the ruby pin F reaches the square notch in the lever it will carry the lever with it. As soon as the lever and pallets have moved a short distance the tooth resting on the pallet will be released and pass across the impulse face I, pushing the lever across, and instead of the ruby pin leading the lever the lever will push the ruby pin, giving the balance an impulse. The tooth giving impulse finally drops off the pallet corner, and another tooth falls upon the locking face of the other pallet. The balance travels round three-quarters of a turn or so and then returns, going through the same series of movements as before, but in the reverse direction, and so on. The angular movement of the lever is limited by the "banking pins" JJ, which are placed in such a position as to hold the lever in the correct position for the ruby pin to enter

the notch properly. The locking faces of the pallets are so
shaped that the pressure of the scape-wheel teeth upon them
tends to draw the lever against the banking pins and hold it
there. There should be enough "draw" to keep the lever to
the pins, but not enough to oppose any serious resistance to
the balance in unlocking the teeth. In spite of the "draw,"
an accidental shake might move the lever and unlock the
wheel. To prevent this there is a safety pin, or guard pin, G,
set upright in the lever, which would come in contact with the
roller edge. To allow this pin to pass while the lever is giving

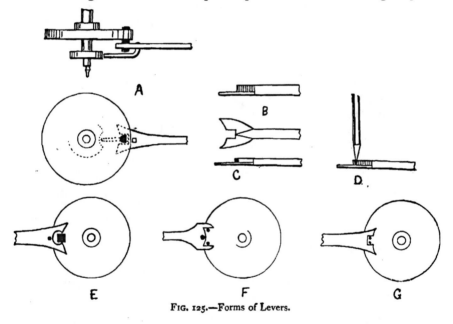

FIG. 125.—Forms of Levers.

impulse, a hollow is cut in the roller edge, opposite the ruby
pin.

This escapement has an ordinary lever and roller action,
but there are many variations. A "double roller" is shown in
Fig. 125 at A. In this there is no hollow in the roller edge,
and no guard pin in the lever; but, instead, the lever has a
"dart" affixed to it, which acts in conjunction with a second
small roller underneath the large one. This is a very safe form,
and much to be preferred to the ordinary "single roller." As

shown in Fig. 125, the "dart" is a bent wire riveted into the lever. In English watches the dart is generally of gold and fixed to the lever by a screw and steady pin. In the best Swiss watches also the dart is fixed in this way, but is of steel and very light. It may be mentioned that the little screw is the smallest made and used in watchwork, and probably the smallest screw used for any purpose.

At B (Fig. 125) a common Swiss form of lever is shown. There is no guard pin, but, instead, the upper part of the lever is filed up to form a dart. When these escapements require the dart advancing to make the action safe, they cause a lot of trouble. There are three methods of doing it. One is to file back the dart point and soft solder a new brass "nose" on it, filing up to shape, as at C; or the dart may be slit with a screw-slitting file, and the end forced forward by a chisel-shaped punch being hammered into the slit, as at D; or the lever may be softened between the pallet-staff hole and the fork, and hammered on its sides to stretch and lengthen it. Before doing this take its measure with the millimetre gauge and see how the stretching progresses. When done polish the lever up again.

Sometimes a square ruby pin is used, as at E, working between the edges of a circular lever notch; or the ruby pin may be replaced by two metal pins, as at F and G. These pins are generally of gold. F is a "two-pin" escapement, in which the guard pin in the lever helps to give impulse by means of the roller notch. G simply has the ruby pin replaced by gold pins. Gold pins get worn into notches, and when so worn should be replaced. They also cut the sides of the lever notch, and necessitate smoothing and polishing the notch with oilstone dust and red-stuff.

Faulty Escapements.—Lever escapements may have many faults, some of construction, some caused by careless bending of banking pins, etc., and some by wear. A most serious fault is "mis-locking." When this fault is present the points of the scape-wheel teeth, instead of falling just on the locking faces of the pallets, as at A (Fig. 126), miss the corner and fall upon the impulse faces, as at B. This is caused by the points of the scape-wheel teeth wearing short, or by the scape or pallet pivot holes wearing wide, allowing the scape wheel and pallets to get too far away from each other. A very little is sufficient to cause a locking escapement to mis-lock. A watch that mis-locks is sure to stop. To test for this, lead the balance rim slowly round with the finger tip, while a little pressure is applied in a forward direction to the scape-wheel teeth with a flat-cut peg point. Watch it give impulse until a tooth "drops." At this moment

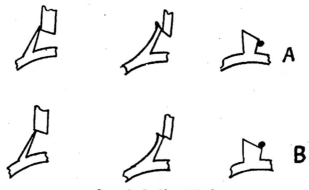

FIG. 126.—Locking of Teeth.

let the balance go free. If it has locked, the scape wheel will remain stationary and draw up the lever smartly to its banking pin. If it has mis-locked the lever will go back and the scape wheel will go forward, giving another impulse in the reverse direction. An escapement should be tried thus before taking the watch to pieces, so that, if faulty, it may be corrected.

American watches, and some English and Swiss, have sight holes in the plate, through which the pallet depth can be easily observed without having to test as above.

To correct a mis-locking escapement, if due to wide pivot holes, bush them; if due to worn teeth, bush the top pallet hole, drawing it a little nearer to the scape wheel, as shown in Fig. 88, p. 83.

Frequently a lever watch is found stopped with the lever on the wrong side of the roller. This is caused by the guard pin being too far from the roller edge and allowing the lever to pass when shaken. To remedy, the guard pin may be bent a little forward, until it will not pass or stick against the edge. A guard pin that will not quite pass, but sticks, is sure to stop the watch, and a watch that is found to stop with the slightest movement—taking out of the pocket to see the time, etc.—very likely has this fault. A slight roughness on the roller edge will also cause the guard pin to catch and stick.

Swiss watches with solid "darts," as in Fig. 125, at B, are more troublesome, and the methods of advancing them are described on p. 104.

The position of the banking pins is often faulty. They should be wide enough apart to allow of a little "banking shake" on either side of the roller, and also wide enough apart to allow the lever to run up to the banking pin a little after a tooth has dropped. This is termed the "run." Only a little "run" is wanted, and only a little shake, but there must be some of both. If the banking pins are too close there is apt to be no run on some teeth, which would stop the watch

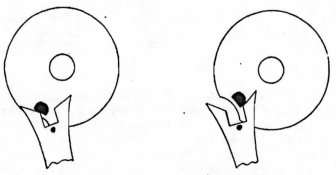

FIGS. 127 and 128.—Faulty Lever and Roller Actions.

instantly, or not sufficient banking shake, which would cause the guard pin to touch the roller edge ; or it might cause the face of the ruby pin to butt against the corner of the lever notch in entering, as in Fig. 127. If too far apart the " run " will be excessive, causing resistance to unlocking and a bad action ; or the ruby pin may butt against the other corner of the notch, as in Fig. 128. A watch that has a bad action when hanging up, with the balance leaning towards the lever, will generally be found faulty in the lever and roller action. If it has a round ruby pin, change it for one with a flat—D shaped—as shown in the figures. To test for these little faults, insert a wedge of pith or cork under the balance rim to keep it steady. Lead it round with the finger tip very slowly, and with tweezers try the shake or freedom of the lever at each point. Any tight spot, caused by fouling, will be found at once.

American watches, and many machine-made English and Swiss also, have adjustable banking pins, which are a great convenience. These are usually screwed into the watch plate, and the pin is eccentric. A slight turn of the screw head serves to move the pin as desired.

The ruby pin should have just a little side play in the lever notch, and not fit it tight. A flatted pin may be made from a round one by taking a piece of brass and filing a groove across its face, deep enough for the pin to lie in with one quarter of its thickness standing above the surface of the brass. Cement the pin in the groove with shellac. Screw the brass in the bench vice, and flat the pin down with a $\frac{3}{0}$ emery buff.

It may be mentioned here that "ruby" pins, like "jewel" holes and pallet stones, may be *glass* in very common watches, are *garnets* in most, and *ruby or sapphire* in a few of the very best. A ruby could not be thus cut with emery buffs, but glass or garnet can.

To cement a ruby pin in a roller, a roller-holder, like Fig. 129, will be wanted. This is to warm the roller upon to run

FIG. 129.—Roller Holder.

the shellac. Holding the roller in or over a flame would probably blue it, and in any case the roller would not remain hot long enough to enable the pin to be set true. The holder is made of stout brass wire 4 in. long, filed smooth, and tapered at one end for about 1 in. At 1 in. from this end a circular piece of brass is driven on tight, to act as a heat retainer. To use it the roller is pushed on the tapered end, the brass block A is held for a moment in the spirit lamp flame and then removed. Enough heat is retained to keep the roller hot. The remains of the old broken pin are then pushed out with a peg. Still holding the roller on the tool, select a new pin to fit, put it in position, and pick up a small piece of flake shellac. Warm the tool again, and apply the corner of the flake of shellac to the back end of the ruby pin until a little melts and runs on it. Work the pin backwards and forwards a little in the hole, and then, while the shellac is half set, set the pin upright with tweezers, and see that the flat is outwards. When cold, cut off the surplus shellac from the pin sides and the roller face and edge with a sharp graver.

In handling a roller—to take it off a balance staff or put it on—use only brass-nosed pliers. Steel pliers will rough its edge and cause the guard pin to catch. A tight roller is best removed with a "roller remover," a useful little tool with a lever action, to draw them off without injury.

Many common watches, especially some English full-plate fusee levers, have pallets not suited to the scape wheel. If the pallets are too small for the wheel they will also be too close together, and the "inside" drop of the teeth will not be sufficient. The scape teeth must have *some* drop on to the pallet locking faces, or else the point of the pallet will scrape down the back of the tooth. With pallets too small for the wheel, there will not be enough drop upon the inside locking face, and the point of the short or "entering" pallet will scrape the backs of the teeth. If the pallets are too large there will not be enough drop on the outside locking face, and the point of the other pallet will foul the teeth backs.

This fault causes a bad action, and can be detected by putting the lever and pallets and scape wheel in a depth tool and adjusting to a correct depth. The action can then be leisurely observed. To remedy, a little must be taken off the faulty pallet corner with a $\frac{3}{0}$ emery buff, or with flour emery and oil on an iron lap in a lathe, or with a "ruby file." It may here be mentioned that English watches have the pallets simply driven tight on to the pallet staff, which is tapered for that purpose, and can be removed by driving the staff out with a punch. The lever is simply pinned on to the pallets with two brass pins, which can be pushed out from underneath. Swiss and American watches have the lever and pallets screwed on to the pallet staff, and can be removed by holding the pallet staff in brass-lined pliers and unscrewing them with the fingers.

English watches generally have "covered" pallets, like Fig. 120, the pallet stones being sunk flush in slits made to receive them. They are cemented in with shellac, and can be re-cemented or moved by warming a pair of pliers and holding them while the stone is adjusted. Swiss and American pallets are nearly always "visible," like Fig. 121. To re-cement or move these stones is easier. Warm a brass plate and clamp the pallets down to it; then operate on the stones. These visible pallets are more fragile and liable to accident than covered ones, and often become loose or chipped. To adjust a pallet stone is a delicate operation. If one has become loose, re-cement it and push it home in its groove. Then try the escapement in the watch frame to see if it locks properly. If the stone wants advancing a little, first take its measure with the millimetre gauge, and the amount it is moved can be determined with accuracy.

If it is desired to try a lever and roller depth in the depth tool there will be a difficulty in getting the balance in. The roller should be taken off and placed on a true turning arbor, and this put in the tool.

To test the "draw" on the pallets, with the tweezers or a peg point, move the lever a little away from one banking pin, but not enough to unlock the wheel. If there is "draw" the tooth point will draw the pallet up again, and the lever will return sharply to the pin. If there is insufficient draw the lever, instead of returning sharply, will move sluggishly or stop still where moved. This should be tried on each pallet, that is, against each banking pin. Want of "draw" causes the guard pin to drag on the roller edge between the beats and makes the action poor. Pallets with no draw can only be properly corrected by a pallet maker. Pallet making is a trade to itself, and is one of the many subdivisions of labour in the making of a watch. Some repairers give "draw' to a pallet by shifting the stone a little aslant, but this upsets other adjustments, and is not to be recommended.

Bent scape-wheel teeth can generally be straightened with tweezers by the eye as a gauge. To make sure, put pallets and wheel in the depth tool and examine the drop before putting the watch together.

Fig. 130.—Shape of a Wheel Seating.

Unlike the other wheels of the watch train, the scape wheel is mounted on a brass collet or seating, instead of being riveted on the pinion. The collet is driven tight on the pinion, and then turned to receive the wheel, as in Fig. 130. A thin riveting edge, A, is left standing above the wheel surface. This is not turned over with a punch, but is burnished over as the wheel runs in the turns or lathe by a fine burnisher with a smooth, rounded point, lubricated with oil. If a scape wheel is found to be loose on its collet, re-burnish it a little.

To turn in a new pallet staff will require no special directions; a piece of steel wire being centred, hardened, tempered blue, turned taper to fit the pallets, polished and pivoted as described on p. 87.

A balance staff is not such a simple job. They are of two kinds (Fig. 131). A shows a rough "brass-collet" staff as obtained from the tool shop, and B a rough "solid" staff. Most repairers use these rough bought staffs, though many make their own. The brass-collet staff A is a piece of hardened and tempered steel ready centred for turning, with a brass collet driven on to it. The solid staff is made of one piece of steel and is soft, requiring hardening and tempering. A brass-collet staff can be made by taking a short length of steel wire, hardening, tempering, and centring it, turning it tapered, and polishing it like a turning arbor, to fit the roller; then drilling a piece of brass for a collet and turning it upon a separate arbor, finally driving it on the polished staff to the correct height.

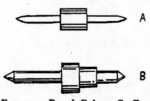

FIG. 131.—Rough Balance Staffs.

A solid staff is made by turning roughly to shape from a piece of soft steel rod, hardening and tempering, then finishing. In any case, the first thing to do is to rough down the collet to size, leaving it just a little too large everywhere. Then to cut the lower part of the staff to length, so that when stood up on the bottom endstone (hole removed) the balance seating can be sighted at the correct height. Then turn and polish the lower part of the staff, tapered to fit the roller, seeing that it also comes at the right height to work with the lever. Turn and polish the bottom pivot to fit the jewel hole and round its end up. The staff can then be reversed and the top-end cut off to length. The total height can be measured by filing a piece of brass to stand between the endstones, with jewel holes removed; or can be measured over the outsides of the holes, with endstones removed, by douzième or millimetre gauge.

The balance seating can next be turned flat and true to

receive the balance *tightly;* the riveting face must be turned hollow to leave a nice thin edge for riveting over the balance, like A (Fig. 130), the hairspring collet fitted a nice sliding fit, and the surplus metal of the staff collet cut off, leaving the upper part of the staff from which to turn the top pivot. This can then be turned, polished, and rounded up. Finally, rivet on the balance. Pass the staff through a hole in the steel stake, so that the collet rests on the stake. Use a crescent-shaped punch (Fig. 132), and tap round and round gently until the riveting edge is well down. Then use a flat-faced punch with central hole to finish, and turn off the riveted face smooth.

FIG. 132.—Crescent or Half - round Riveting Punch.

True the balance in the way described on p. 84 and "poise" it. If a watch is to keep time, the balance must be in perfect poise, that is, it must have no heavy part and show no tendency to settle at any one point when placed in a pair of calipers. A balance poising tool consists of a pair of parallel straightedges that can be adjusted to suit various balance staffs. The pivots are rested upon these, and any heavy part will at once settle at the bottom. To poise a plain gold or steel balance, file a little off the inside edge of the rim at the heavy part. A compensation balance can be poised by its quarter screws, or if out too much for that, small washers (sold for the purpose) may be put under the screw heads, or screws can be slightly reduced.

A balance staff has cone-shaped pivots for strength. These can be turned by a thin graver with a rounded point, rubbed round on the oilstone. The polisher must be curved in section to suit them, as in Fig. 133, C. They should be smoothed with oilstone dust until the ends just go in the jewel holes, and polished with red-stuff until the ends come well through, as at A (Fig. 133). Although called "cone" pivots, the part that goes in the jewel hole must be straight; only the shoulder is curved, for strength. A pivot of this shape will naturally draw the oil away from its point and let it spread up the staff. To prevent this, cone pivots are "back cut." An enlarged "back cut" pivot is shown at B.

In the turns a balance staff is manipulated exactly as

described in turning pinions. In a watch lathe the staff is roughed out in split chucks, then cemented in the shellac chuck (Fig. 100, p. 89), and the entire lower part finished ; then taken out, measured, and cut off to total length, and reversed

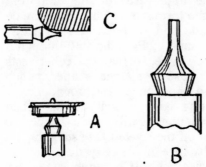

FIG. 133.—Polishing Cone Pivots.

in the chuck, the balance seating, etc., and top pivot being done last. The whole is done in two chuckings.

Fig. 134 shows an undersprung and an oversprung staff. A = seating for hairspring collet. B = balance seating.

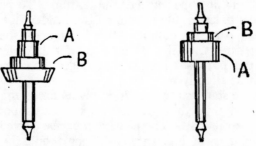

FIG. 134.—Undersprung and Oversprung Balance Staffs.

Bent balance-staff pivots are straightened with brass-nosed pliers. Sometimes, when a watch has had a blow, the end of one pivot is riveted like a miniature mushroom against the face of the endstone, as at A, Fig. 135, and cannot be withdrawn from the jewel hole. In such a case something is bound to smash. The pivot must be pulled out by main force and may part, or the jewel hole may break. A broken pivot altered to fit round, by or over a cam. It may even develop into a small rack and pinion motion. Whatever its shape, it has only one purpose : to regulate the length of the balance spring. The mechanism is to permit very small movements to be made and for the index to be held firmly in any position.

The index will swell into a bright circle surrounding the top balance pivot endstone and from beyond will project a small finger overhanging the spring. With a flat spring the projection will be long, extending slightly past the outside coil ; for a Breguet spring the projection will only reach the overcoil. Under the projection and pointing downwards are two little pins, the curb pins, which enclose one coil of the spring without gripping it. The curb pins permit the spring a slight sideways movement, just curbing its action. This is the count-point of the spring. Movement of the curb pins towards the outer end of the spring will have the effect of lengthening it—the watch will lose ; movement towards the inner portion of the spring will shorten it—the watch will gain.

Sometimes after a sudden jolt a wristlet watch will begin to race and will gain forty-five minutes a day. This indicates that rough treatment has caused the second coil of a flat balance spring to catch up on the index curb pins. Let the repair man put it right ; do not tamper with it. In addition to releasing the caught up coil the curb pins need adjustment.

"SHOCKPROOF" ESCAPEMENTS AND MAGNETISM

THE balance staff presents many problems, not the least being the fragility and hardness of the pivots ; their liability to bend or break. Thick pivots interfere with good time-keeping, thin pivots break easily ; so a compromise is found.

In a small watch a balance staff may be only three six-teenths of an inch high, yet it must be as complete and perfect as a large staff. It must have at least six main diameters with true and square shoulders, accurate tapers and nicely rounded-off pivots. Mathematical balance for the balance wheel can never be obtained on a staff that is out of true. All the parts that fit on the staff: the large roller, small roller, balance wheel and balance spring collet, must each be a good push fit on its own seating, and each, also, must be at the correct distance in relation to the other components of the escapement.

Examine a small movement and observe the narrow clearance in height allowed between the balance wheel and the centre wheel which overhangs it, and the lever-bridge which is beneath. Unless the heights were exact no end of trouble would arise from the balance wheel fouling something else, not enough to stop it, but occasionally when turned from one position to another. Mysteries of erratic timekeeping can sometimes be solved by examination for this fault.

Apart from keyless work troubles and broken mainsprings, ninety per cent. of *breakages* in watches are fractures of the balance staff pivots. This is understood when are considered the fine dimensions of the pivots, the length of the staff between them and weight of the rollers, balance and collet it carries. When a watch receives a blow on the side or edge of the case, the mass of the balance resists the sideways action at the same time as the pivots are attempting to carry it forward. If the shock is slight no damage may result; if the shock is heavy the least that can happen is one or both of the pivots will bend; the most is that one or both will break. A watch may run indefinitely with a slightly bent balance pivot, but no precise timing will be possible. If a pivot is snapped off, obviously the watch will stop.

Balance Staff.

To minimize accidental shocks of this sort certain modifications of the escapement have been invented from time to time. Breguet, the inventor of the overcoil spring, devised more than one form of resilient bearing for the top balance pivot in his watches. At that time they were known as parachute escapements. Improvement in design caused these parachute escapements to be soon abandoned; the idea has recently been revived and more than one patented attachment for the purpose of taking care of the balance pivot when used roughly are on watches now in wear. Watches fitted with these shock absorber types of jewel hole assemblies have been hailed as " unbreakable."

Can a watch be " unbreakable " ? Test it this way. Take an ordinary building brick and lay it carefully on a paving

20

stone. Now strike the brick a hard blow with a sledge hammer. What is the result? But a brick is not made to stand a sledge-hammer blow. Nor is a watch made to stand anything but careful handling. It is a delicate scientific instrument, not a missile. If a watch is fitted with a protective device for the pivots, it is so that closer timekeeping may be secured by the use of fine pivots. Misguided jewellers' salesmen may wrongly proclaim a watch as " unbreakable." Please ignore them ; nothing is unbreakable. A watch may be sturdy enough to stand the roughest of ordinary wear, but that is quite a different thing and all that should be claimed for any watch. Every blow a watch gets will cause some difference to its timing ; blows displace delicate adjustments even if they do not bend or break delicate parts.

The theory of a resilient jewel hole is that the jewel will have a spring-held setting with a slight allowance for side and end movement. In a popular pattern the jewel assembly is seated on a cone bearing and is held in place by light spring pressure from the top. After movement in any direction the assembly is returned to its central position. The best feature of the device is that movement of the jewel assembly in any direction permits a strong shoulder of the staff to come up against a part of the balance cock or bottom plate. This strong shoulder takes the shock instead of the pivot and does it without damage.

In another pattern the jewel hole setting floats centrally in a cleverly designed small spiral spring. The arrangement for taking the shocks on a shoulder or thickened part of the staff are on the same principle as the previously mentioned fitting. Yet another method is applied to the balance wheel itself. Instead of the wheel having a diametrical arm, the arms are formed of two curves rather like an exaggerated S. This gives a certain springness to the balance which swings surrounded by a retaining ring. The effect of a knock is to displace the balance wheel, but as it has a small amount of freedom, it hits the retaining ring before the movement is felt by the pivots. Vertical movement is limited also. This is a clever and straightforward attempt at the solution of a difficult problem, except that any great amount of tapping by the wheel on the retaining ring must surely affect the timekeeping and should the balance get out of true repoising it is a tricky job to put right.

Magnetism is as fatal to a watch as cramp is to a swimmer. The effects are similar. When a watch is mildly magnetized the steel parts—springs, screws, winding shaft and winding wheels—influence the balance spring. Strong magnetic influence will grip the vitals of the watch and put it completely out of action. The prevention is to keep the watch away from magnetic fields ; the remedy is for the repairer to remove the trouble with a de-magnetizer, an electric machine specially devised for the purpose.

Cylinder Escapement.—This form of escapement is going out of use rapidly. Probably in the near future it will only be found in some Swiss ladies' watches, for which it answers fairly well.

Fig. 136 shows the action and the different parts. A = cylinder, B = scape wheel. It consists of a scape wheel with fifteen peculiar teeth and a hollow steel cylinder, a part of which is cut away. Each tooth of the scape wheel is mounted upon an upright stem, standing up from the scape-wheel rim. On each tooth is an inclined plane or nearly plane curve. The central part of the cylinder, where the scape-wheel teeth engage it, is cut away, leaving just a little more than half the circle. The wheel and cylinder are planted so that the centres of the inclines on the scape-wheel teeth pass through the centre of the circle of the cylinder. Suppose a tooth rests inside the cylinder as in the figure at A. The point of the tooth presses against the inside surface of the cylinder shell and the wheel cannot advance. As soon as the balance and cylinder turns in the direction of the arrow the tooth point will be released, as at C, and the scape wheel will advance, the inclined face of the tooth giving an impulse to the cylinder edge or " lip." When the tooth has quite passed, the next tooth will fall against the outside surface of the cylinder and lock the wheel once more. The cylinder returning, the point of this tooth will be released and it will give impulse as at D, until its point drops against the inside surface of the cylinder shell, and so on.

This is really a very simple escapement; but because it is somewhat hidden in the watch its action is much misunderstood, and is a great mystery to many beginners. The cylinder must be of such a size that it passes between two teeth of the scape wheel with a little to spare, as shown at E. Also it must be large enough for a tooth to be completely inside it and have a little shake, as at A. The amount of shake gives the "drop" of the teeth. Thus a cylinder too large for the wheel will have no outside "drop," while one too small will

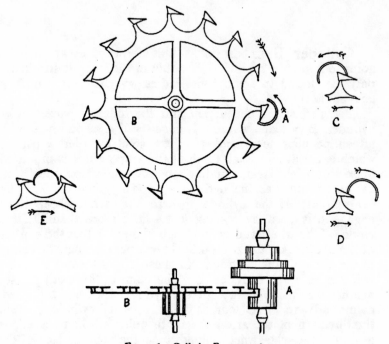

FIG. 136.—Cylinder Escapement.

give no "drop" to the teeth inside. In a correct escapement the inside and outside drop should be equal. A watch that suffers from too little drop, either inside or outside the cylinder, may be made correct enough to go by filing the least trifle off the points of the teeth, shortening them a little. They must all be done equally, and not one missed.

The depth of the cylinder and scape wheel is not a very particular one within small limits. But if too shallow it will "run through." This is caused by the points of the teeth missing the surface of the cylinder and getting past its edges, thus giving impulse as soon as they drop. A watch with this fault will tick rapidly and unevenly. Too deep a depth is shown by a want of shake of the teeth inside the cylinder and by the balance having to turn a long way round to unlock the teeth points.

The depth can be observed by leading the balance rim round with the finger tip and watching the motion of the scape wheel carefully. In a correct depth the scape wheel will advance as it gives impulse, and then "drop" on to the surface of the cylinder. At this point reverse the motion of the balance. The wheel should stand still, locked, while the balance is moved a few degrees, and then commence to move forward and give impulse again. If the depth is too shallow the wheel will go forward again immediately the motion of the balance is reversed, there being no pause at all. If too deep the balance will have to be reversed a considerable distance before the wheel starts again.

In nearly every cylinder watch there is a means of regulating the depth. There will be found a cock or "chariot," like Fig.

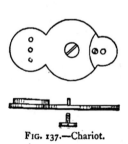

Fig. 137.—Chariot.

137, under the plate. In this the bottom jewel hole of the cylinder is fixed. Into its other end the balance-cock screw and steady pins go. Therefore if this chariot is advanced, or backed a little, the whole cylinder moves towards or away from the scape wheel. To advance it, bend its steady pins backward and draw the screw hole a little oval with a file. To back it, bend the steady pins forward. It may be necessary, to enable the chariot to be moved, to trim a little off some of its edges or off the plate by filing.

Another important point in this escapement is the relative height of the cylinder and scape wheel. The scape wheel should pass through the "passage" cut for it in the cylinder without touching either the top or the bottom of it, as shown in Fig. 136. When it touches either the top or the bottom badly the watch will stop; and while going the scape wheel will be seen to "dance" at each beat. When held to the ear, first dial up, then dial down, a knocking or scraping will be heard. Too much endshake of the cylinder or scape wheel will cause fouling here. The endshake of both wheel and cylinder should be very little, and be equal. They will then not alter their relative positions when the watch is turned over. A cylinder that is too high can be got down by putting a small slip of paper under the chariot. One too low is not quite so easily got up. If the bottom endstone is set in a brass slip, the cylinder can be raised a little by breaking the endstone out, chamfering its setting out, and fitting a loose endstone as thick as it will take. This saves the room that was wasted between the endstone and the jewel hole. Or the face of the chariot may be filed so that it will lie closer to the under side of the plate.

In the rim of the balance there will be found a little pin. This is to prevent the balance making more than half a revolution in either direction from its point of rest. It should come in contact with a fixed pin in the back of the balance cock. If the pin in the rim is too short or is lost, and the balance makes more than half a turn, it will be caught by the scape wheel teeth and held firmly from returning. Or if the pin is in the wrong place, allowing the balance more than half a turn in one direction, the same thing will occur. In some watches the pin in the rim banks against the fourth pinion instead of a pin the back of the cock. The pin in the rim should be in such a position that, when the watch is in beat and the balance at rest, it is exactly half a turn from the pin it banks against.

The cylinder itself is a thin steel tube. Its ends are closed with turned steel plugs driven in tight. Upon these plugs the pivots are turned. Upon the outside of the tube a brass collet is driven, and is turned to form a seating for the balance and the hairspring. Cylinders in the rough—that is, with the tube finished and polished, the plugs fitted, and the collet on—can be purchased for a few pence. They require cutting down to the correct length, pivoting, and the balance and hairspring fitting.

To turn in a new cylinder is not such a difficult job as many beginners think. In a watch lathe the cylinder can be held in the cone cement chuck, like a balance staff, and the lower portion turned and pivoted, the cylinder body being filled with shellac to stiffen it. It can then be taken out, measured for correct height, and reversed in the chuck for finishing the balance seating and top pivot. In the turns the cylinder must also be filled with shellac, as in a lathe. To do this, place a flake of shellac inside and warm it gently over the lamp flame—not *in* the flame, or it will be softened. A brass

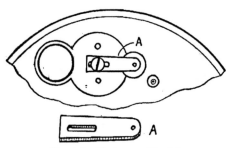

FIG. 138.—Temporary Chariot.

ferrule, with a plain circular hole in its centre, is cemented on the cylinder with shellac. Whether this ferrule runs exactly true or not does not matter. The pivots are turned and polished, etc., as before described on p. 86.

The principal difficulty is to get all the heights right. There are cylinder height tools sold, but they are not really necessary. The simplest process is to remove the under

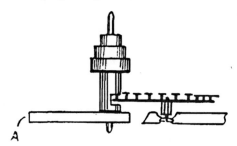

FIG. 139.—Adjusting Height of a New Cylinder.

chariot from the plate, and in its place screw the brass piece A, Fig. 138. This piece can be adjusted so that the pivot hole comes about the right place. The hole should be large enough to take the cylinder plug and allow the body to stand upon it, as in Fig. 139. It can then be seen if the passage in the cylinder comes at the right height for the scape wheel. The bottom of the cylinder body must be turned away until it is right, and a *little more*. Then turn the bottom pivot, leaving it full long, shorten it down until when in its jewel hole it is right height. Next sight the level for the balance so as to just free the scape cock. Turn this and the shoulder for the hairspring collet. Finally, measure the total height of cylinder with douzième or millimetre gauge, over the outsides of the jewel holes with endstones removed, and turn the top pivot. In riveting the balance on, care must be taken to see that the banking pin comes in the right place, as explained on p. 117.

A bent cylinder can rarely be straightened without breaking, but it may be tried with brass-lined pliers, springing it carefully. Bent pivots can generally be managed. A broken pivot is replaced by knocking out the cylinder plug and fitting a new one, turning a fresh pivot upon it. Cylinder plugs can be purchased ready made, but are not necessary if the workman has a lathe. They are so easily made.

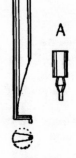

FIG. 140.—Cylinder plug Punch.

To knock out a cylinder plug is sometimes very difficult. The cylinder is first rested upon the brass collet, so as to give a good seat, and the punch shown in Fig. 140 used to drive it out. If the plug is tight the cylinder may be driven through the collet, instead of the plug moving. In such a case, rest the cylinder end over a hole in the stake which the plug will go through and the cylinder will not, and try again. Or a stake with coned holes may be used, as in Fig. 141, which allows the plug to be just started. When a plug has been knocked out just a little and is still tight, it need not be further touched, but the new pivot may be turned upon it. A plug that cannot be moved by any means may, in a watch lathe, be easily drilled, and a pivot put in as if it were a pinion. A " plug " is shown in Fig. 140 at A.

A scape wheel bent out of flat may be made true by bending with tweezers, springing it carefully in the desired direction. Broken teeth cannot be properly replaced, but require a new scape wheel. New wheels cost a few pence each, and can be purchased the exact size required, ready to put on the scape pinion. The scape wheels are not riveted on the pinions, but only driven on tight a good fit, and may be easily removed by punching the pinion through.

A new scape wheel will require opening out in the centre to fit on the pinion. As bought they are too hard to broach.

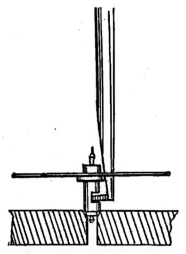

FIG. 141.—Removing a Cylinder Plug.

The centre may be softened by heating a tapered brass wire and inserting into the hole. Or the pinion may be held in a split chuck in the lathe and turned down to fit the wheel.

A broken scape pivot can be put in with a watch lathe by first knocking off the wheel, then putting the pinion in a split chuck and drilling. A pinion with both pivots off may be drilled right through and a piece of steel driven in, a pivot being turned on either side. This job in the turns is a nearly impossible one.

The tops of the scape-wheel teeth sometimes run foul of the scape cock. To free them, the cock must be cemented flat on a piece of sheet brass with shellac and put in the mandrel ; or it may be cemented on to a flat-faced cementing chuck in a watch lathe, and the passing hollow turned out a little more.

Old English cylinder watches have brass plugs to the cylinder, in the centre of which steel inner plugs are driven to form the pivots upon. These watches also often have brass scape wheels. This causes the cylinder sides to wear badly. A new cylinder to one of these watches is not a very formidable job. A piece of steel rod is drilled, and smoothed inside by a brass rod and oilstone dust and red-stuff. The plugs are fitted and the collet driven on. The outside is then polished and the body cut open.

Verge Escapement.—This escapement requires very little explanation. It is now nearly obsolete, and, it is hoped, will be quite so in a few more years.

The scape wheel A, Fig. 142, has thirteen or fifteen saw-

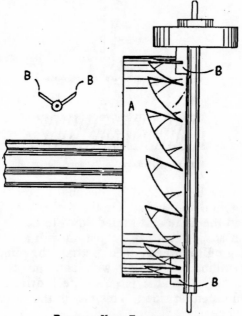

FIG. 142.—Verge Escapement.

shaped teeth. It runs vertically in the watch, the back pivot running in a "follower," and the front or "inside" pivot in a brass slip in the potance face (see Fig. 71, p. 66). The teeth points act alternately on the pallets BB, which are flat steel "flags" projecting from the slender verge body. In this escapement the balance is never free from the scape wheel; when the teeth are not driving the pallets, the pallets are causing the teeth to "recoil." Therefore any variation in the motive power directly affects the time of the watch, and a "fusee" is a necessity. Also a stronger mainspring will cause the watch to gain, or a weaker one will make it lose, which is not the case with either a lever or a cylinder watch.

It may have, and generally has, many faults. The depth may be too shallow, in which case the "drop" on both pallets will be excessive, and the balance will perform very short arcs, causing the watch to gain. If too deep the teeth points will catch on the verge body, or there will be no "drop," and the watch will stop. This depth should be set as deep as possible for it to go and not catch. It can be regulated by driving the "follower" in a little, and so getting the scape wheel nearer to the verge.

The drop on the pallets can be adjusted by sliding the slip which carries the inside scape pivot a little to one side, until the drop on each pallet is equal.

A frequent cause of trouble is wide worn pivot holes. The inside scape pivot hole gets worn badly. This should never be passed, but must be bushed a good fit. To bush this hole, do not remove the slip, but bush it in position. If at any time the slip has to be moved to adjust the depth or for any other purpose, it is best to make a scratch across its face so that the amount of movement can be seen.

Verge pivot holes are frequently "dead" holes in brass. To fit new ones, knock out the old brass plug in cock or potance, or drill it through and broach it out. Then file up a good flat-ended round brass pin to fit tight with the flat face flush inside. Drill it up with a flat-ended drill to a sufficient depth, a little eccentric, not exactly in the centre of the brass. Then put in the verge and turn the eccentric brass plug round until the verge is upright. Cut it off level and rivet in.

To properly re-turn and polish an inside scape pivot that is badly worn, the scape wheel must be knocked off. When riveted on again it must be topped true by revolving rapidly in the turns and holding a pivot file to the teeth points until it just marks all of them. They must then be filed up to good points by operating on the *curved backs* only with a fine crossing file, the wheel being held in a pin vice by its pinion.

Verge pallets that are cut into holes by the teeth may be

smoothed down and polished with oilstone dust on a flat steel polisher, followed by red-stuff.

A little fault often found in these escapements is that the top of the scape wheel touches the under side of the brass collet of the hairspring.

In a watch lathe a new verge is put in by cementing in a cone chuck to turn the collet and pivots. In the turns, the slender verge body is stiffened by cutting a short piece of watch peg as in Fig. 143. The verge is placed in the hollow and the loose piece fitted in, pushed tightly into a solid brass ferrule and shellaced firm. It can then be turned and pivoted. The pallets must be shortened off by filing to make the escapement drop off properly, and the backs bevelled off to a knife edge.

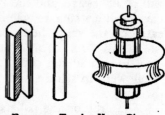

FIG. 143.—Turning Verge Pivots.

A verge balance has an upright banking pin in its rim, which banks against two fixed studs, or two shoulders in the balance cock. When in beat, and the balance at rest, this pin should allow the balance to turn an equal distance in either direction.

Duplex Escapement.—This is not now made, with the exception of " Waterbury " watches, which still have it. Generally a good old English watch with this escapement pays to convert to a lever.

Fig. 144 shows the action. The scape wheel A has two sets of fifteen teeth. One set are long and pointed (D), the other are short and upright (C). The teeth D are for locking the wheel, and rest upon a ruby roller, E, on the lower part of the balance staff. This roller has a vertical slit, which just allows the teeth points to pass as the balance comes round. The upright teeth C are to give impulse, and, just as the ruby roller allows a long tooth to pass, one of the teeth C gives an impulse to the point of the long pallet B on the balance staff.

The pallet B is fitted on to the staff like the roller in a lever escapement, and can be turned round as desired to adjust the drop. The ruby roller E is a tube of ruby slit down one side, and is slipped a sliding fit on to the lower part of the balance staff and cemented with shellac. A small brass cap, F, is pushed on after it and cemented also.

With wear the points of the teeth D shorten, and this, combined with wide pivot holes and worn pivots, causes the teeth to slip by the ruby roller. The remedy for this is to repolish

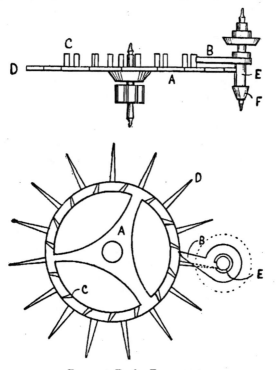

FIG. 144.—Duplex Escapement.

the pivots and fit new jewel holes, drawing the jewelling a little if necessary. The drop of the short teeth upon the pallet B should be quite perceptible on each tooth, and must be carefully noted by leading the balance round with the finger tip until a long tooth escapes and watching the amount the wheel moves before a short tooth falls upon the pallet. To give more or less drop, the pallet B can be turned round.

In old watches the long teeth points wear off, and the short teeth cut until they nearly miss the impulse pallet. Such a watch wants a new scape wheel. A broken balance staff and a cracked ruby roller are also the frequent results of a fall. Any of these repairs are troublesome and expensive, and, if required, it will generally be found best to at once convert the watch into a lever.

In Fig. 144 the point of a long tooth is shown just about to escape from the slit in the ruby roller, and a short tooth is about to "drop" on the pallet B and give impulse. No impulse is given on the return vibration.

Chronometer Escapement.—This was invented for marine timekeepers, to keep Greenwich time on board ship for the purpose of ascertaining the longitude. It is a very delicate escapement, but is capable of keeping correct time for a long period without much variation. It is sometimes found in pocket watches, and, in the hands of careful wearers, is fairly successful. Still, for such purposes, it cannot be said to equal a fine adjusted lever.

Fig. 145 shows its arrangement, as made in England. The scape wheel A has fifteen teeth of the shape shown. The wheel is locked by the teeth falling upon a locking stone (ruby), C, set upright in a spring detent, B, which is screwed to the watch plate. The balance staff has two rollers upon it. The small one has a ruby pallet, E, projecting from it, which, as the balance comes round in the direction of the arrow, will lift the point of the detent B and let the tooth H escape. The tooth I will then fall upon the pallet G in the large roller F and give the balance an impulse. This escapement, like the duplex, only gives an impulse at each alternate vibration of the balance. On the return vibration the pallet E only raises the thin gold passing spring D, and does not move the detent itself.

In this escapement the balance is more free than in any other, and it requires no oil upon the scape-wheel teeth or on the detent point.

Both rollers can be moved upon the balance staff for adjustment. As in the duplex, as soon as the detent is raised and a tooth, H, escapes, a tooth, I, should have a little "drop" on to the pallet G. When the wheel is locked by the detent, the large roller F should be just free of the points of the two scape-wheel teeth between which it revolves.

The detent banks up against a screw stud, and its exact position can be regulated. To be correct, the gold spring should point exactly to the balance centre. The slightest trace of oil upon the point of the banking screw, or on the detent point, where the gold spring rests upon it, will make these parts sticky and cause great irregularities of rate.

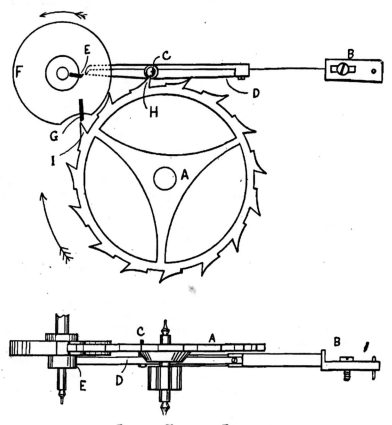

FIG. 145.—Chronometer Escapement.

Broken pallet jewels can be replaced by cementing with shellac. A detent locking stone is half round, as at A (Fig. 146), and should be fitted into its pipe by filing up a half-round brass pin, B, to fill up the interior. This pin should not be wedged in tight, or the jewel will break, but should be an easy fit and cemented altogether with shellac. Warmed pliers are used to hold the " pipe " and run the shellac.

A new gold spring may be made from any scrap of gold wire by hammering out and filing up. Some are riveted and others are screwed to the detent. A screwed spring should have an oval screw hole for adjustment. The point should be square, and the edge clean, burnished, and free from burrs.

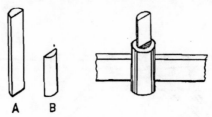

Fig. 146.—Fitting a Detent Locking Stone.

The gold spring should not be of equal thickness throughout, but the point may be thickest and the part near to its fixed end the thinnest, so that nearly all its bending is done near to where it is fixed.

A detent is a delicate thing to make, perhaps the most delicate part found in watchwork. The greatest care must be used in handling one. It is filed up from the solid, foot, spring, and body all in one piece. In making a new one, a length of steel is taken, the position of the foot screw-hole marked, drilled, and filed oval for adjustment, then the hole for the locking stone drilled and a brass pin fitted for trial. It is then roughly shaped up, hardened and tempered, the spring portion being left fairly thick until last. It is now worked down to size all over, and its extreme point softened and turned round for the gold spring to lie upon. When adjusted for length accurately so as to lock the scape wheel in the exact position, the steady pin hole is drilled in its foot and the spring thinned, first by filing, then by polishing down with oilstone dust and oil on a flat polisher until it is thin enough. Finally, it is polished with red-stuff and oil.

Detents are occasionally seen that have broken and been repaired. In nearly every case, though the watch has been made to "tick" by so doing, its timekeeping qualities have suffered badly. In fact, it is no longer an instrument of precision. The only remedy for a broken or damaged detent is to make a new one.

In putting this escapement together, care should be taken to see that the edge of the gold spring does not touch the scape wheel or the banking screw, and that the detent point or the gold spring point does not touch the flat of the large roller.

Some foreign chronometers have a pivoted instead of a spring detent. In these the detent pivots run in jewel holes, and it is brought against its banking screw by a spiral spring like a hairspring, or sometimes by a thin flat spring generally made of gold. Such escapements are not so good as those on the English plan, being more affected by thickening of oil, and not so certain to go together right after cleaning.

CPSIA information can be obtained at www.ICGtesting.com
Printed in the USA
LVOW101149041212

309974LV00001B/367/P